THE HILL TRUTH ABOUT STEROIDS

Everything You Need To Know About Anabolic Steroids

Written by
Jack Osbourne

CONTENTS

INTRODUCTION 1
CHAPTER 1. WHAT ARE STEROIDS? 2
CHAPTER 2. BENEFITS OF ANABOLIC STEROIDS 12
CHAPTER 3. DRAWBACKS OF ANABOLIC STEROIDS 18
CHAPTER 4. HOW TO USE STEROIDS 22
CHAPTER 5. STACKING ANABOLIC STEROIDS 28
CHAPTER 6. WHAT YOU WILL NEED 31
CHAPTER 7. MAXIMIZING YOUR GAINS 36
CHAPTER 8. DIARY OF A USER 43
CONCLUSION 51

INTRODUCTION

Thank you for taking the time to download this book: The Hidden Truth About Steroids.

This book covers the topic of anabolic steroids, and will teach you about the mislead facts, potential side effects, different cycle and the importance of eating the right foods while using them. This book is neither in favor of nor opposed to the use of anabolic steroid, per se. On the contrary, it will provide you with a real-world, clear-eyed look at anabolic steroids, as well as the pros and cons that are associated with their use. Although illegal for most performance-enhancing uses, the use of anabolic steroids by those wishing to perform at their highest athletic levels is as popular as ever.

Ultimately, it is up to each reader to weigh the pros and cons of using anabolic steroids and decide whether they are for you. But, this book will provide all of the information you may need to help you make an informed decision. Even if you have no interest in using anabolic steroids, this book will provide you with an in-depth look at anabolic steroids, the risks associated with anabolic steroids, and how athlete as well as normal people like yourself use anabolic steroids to increase their overall quality of life.

At the completion of this book you will have a good understanding of anabolic steroids and the potential risks and benefits.

Once again, thanks for downloading this book, I hope you find it to be helpful!

Chapter 1.

What Are Steroids?

So, what are anabolic steroids? Anabolic steroids, more properly referred to as anabolic–androgenic steroids (AAS), are simply synthetic replacements or derivatives of testosterone and nor-testosterone. This means that they are synthetic versions of the same hormones produced by your body. First synthesized by scientists in the 1930's, this compound was first used to aid men who were unable to produce enough of their own testosterone to promote normal growth and sexual development.

Then, during World War II, it was found that this compound could be used to help soldiers suffering from malnutrition gain weight back and improve their performance on the battle field.

Later, after the war, athletes discovered this compound and began to use it to enhance their performance in competition. At the 1956 Olympics, this compound was credited for giving Soviet athletes the edge they needed to perform so well. After hearing that Soviet athletes were using this compound to dramatically improve their performance, an American physician named John Ziegler then developed a more selective form of the compound, an oral anabolic steroid called Dianabol, and distributed it to athletes. He quickly discovered that this greatly increased muscle mass and strength in the athletes who took the oral steroid.

After that, up until the early 1970s, the use of steroids exploded in popularity among Olympic athletes, as well as professional and college athletes. Finally, in 1975, the International Olympic

Committee, (or IOC), finally took action against steroids, banning their use for all athletes performing in the Olympics.

In spite of this ban, sales continued to spike over the following years on the black market. In 1988, the United States government passed the first major piece of Federal legislation against the use of steroids. Introduced as part of the Anti-Drug Abuse Act, this legislation stiffened the penalties for selling and possessing steroids. Only two years later, Congress then passed the Anabolic Steroid Enforcement Act of 1990, which listed certain anabolic steroids as schedule III drugs on the controlled substances list. Previously, steroids had not been scheduled and were controlled only by state laws. In spite of this, the illegal sale of steroids is still prevalent and many researchers have found that the use of steroids by adolescents and adults is still quite common.

Today, the only legitimate use of anabolic steroids is for prescribed medical purposes. For example, a pituitary malfunction in boys may cause his doctor to prescribe some sort of steroid to aid in growth as would the absence of testicles in men, (perhaps due to removal because of testicular cancer

Also, steroids are frequently given to men and women who have been in a coma for a prolonged period to assist in preventing or treating muscle degeneration.

They can also be prescribed to cancer patients experiencing muscle degeneration due to chemotherapy.

In all of these medical cases, doctors only prescribe steroids at the same doses as the natural hormones would be produced by the body under normal circumstances.

Although all steroids have the same basic chemical structure, small alterations can produce wildly different effects when it comes to anabolic and androgenic activity. Anabolic activity is the name of a steroid's ability to increase skeletal muscle growth, and androgenic activity simply refers to the ability of the compound to induce the development of male sexual characteristics, such as a deep voice, facial hair, Adam's Apple, etc.

Even if all of the exact mechanisms that anabolic steroids use to cause their effects are not fully understood, all anabolic steroids will increase muscle mass to some extent. Many scientists believe that steroids work by binding to the androgen receptor, (or AR), and activating protein synthesis. This synthesis of protein allows for a dramatic increase in muscle tissue over a comparatively short period of time. These are one type of anabolic steroids.

Another type of steroid either binds very little to the androgen receptors or not at all. These particular steroids most likely act by inhibiting the effects that glucocorticoids exert upon muscle tissue. They do this by limiting glucocorticoids from producing more glutamine synthetase, which normally breaks down muscle tissue. This is known as an anti-catabolic activity. Glucocorticoids have also been found to either influence or outright cause osteoporosis, so inhibiting the effects of these glucocorticoids has been found to be a relatively effective treatment for osteoporosis. This type of steroid is considered by many to be more effective at promoting muscle growth than the first type.

While it is still not 100% understood how exactly anabolic steroids exert their influence, it is these two mechanisms that are given most of the credit.

In spite of their use by athletes no longer being permitted and the regulations that have been set up around them, anabolic steroids have many valuable medical uses, even today. These medical uses can be broadly grouped into three types: anabolic, androgenic, and other uses.

Anabolic Uses

Prevention and treatment of osteoporosis in women after menopause.

Bone marrow stimulation. Although no longer used as commonly as a treatment for hypoplastic anemia, particularly aplastic anemia, for decades anabolic steroids were used as one of the main treatment therapies for these ailments.

Stimulation of growth. Many doctors find anabolic steroids to be an effective part of the treatment plan for children unable to grow normally.

Cancer and AIDS. Due to the appetite stimulation qualities attributed to anabolic steroids, as well as the promotion of muscle mass growth, anabolic steroids have been found to be an effective treatment for those suffering from chronic wasting diseases.

Androgenic Uses

Androgen replacement therapy for men suffering from low levels of testosterone. This is mainly given to men suffering from a decreased libido.

Hormone therapy for transgender men. Due to their ability to produce male secondary sexual characteristics, such as a deep voice and facial hair, they are often given to transgender men to aid in their conversion.

Induction of male puberty. In cases of an extreme delay in puberty for boys, physicians will occasionally prescribe anabolic steroids to promote increased height, weight, and mass in the boys.

Other Uses:

In spite of certain side effects, anabolic steroids can be used to treat breast cancer in women, although this has mostly fallen out of favor with most physicians.

Hormone therapy for women and transgender women. In postmenopausal women and transgender women, the use of anabolic steroids has been found to increase energy, sexual appetite, and overall quality of life.

Male contraception. In the form of Testosterone Enanthate, anabolic steroids have the potential for use as a safe, dependable, and reversible contraceptive for men.

Now that we have taken a brief look at some of the medical uses of anabolic steroids let's examine how they are mainly used in daily life. Most of the people taking steroids are not professional athletes. Around 1% of the United States population is believed to have used anabolic steroids at some point. In the United States, the vast majority of users are middle-class heterosexual men around the age of 25. This group mainly consists of non -professional bodybuilders and non-athletes who use the compounds for cosmetic purposes. Studies in the US have found that the most users of anabolic steroids have a higher level of academic achievement, a higher median income, and a higher employment rate than similar men in their age group. Most users of anabolic steroids tend to perform more research on the drugs they are taking than users of other illegal substances, although much of the information comes not from physicians or medical journals, but rather from word of mouth, the internet, fitness magazines and other unreliable sources.

Many users of anabolic steroids are happy and successful people using the compound safely to improve their daily lives. The overwhelmingly negative portrayal of anabolic steroids in the media does not match their personal experiences with the drug. And, due to its illegal nature, most users of anabolic steroids do not disclose their use to their physicians. In addition to the illegal nature of anabolic steroids, a lack of trust in their physicians' knowledge about the non-medical use of those compounds can also cause users not to disclose their usage to their doctors. In spite of the lack of reliable sources about using anabolic steroids for non-medical purposes and the lack of trust in their personal physicians when it comes to anabolic steroid use, most users of anabolic steroids feel that their use is safe in moderation.

Male and female professional athletes from many different kinds of sports have used anabolic steroids either to give them a competitive edge or to help them recover faster from an injury. These sports include bodybuilding, competitive weightlifting, mixed martial arts, professional football, boxing, etc. While the use of anabolic steroids and many other kinds of performance-enhancing compounds is banned by nearly every professional sport's governing body, this does not prevent adolescents who desire to reach the professional level of their chosen sport from using anabolic steroids to aid in their training. Generally speaking, males are more likely to use anabolic steroids than females and athletes are more likely to use them than non-athletes.

Now that we have discovered who is using anabolic steroids let's take a brief look at what sort of steroids are currently on the market and what differentiates them.

One of the types of AAS already discussed includes Dianabol. The chemical name for this compound is methandrostenolone, but may also be called D-Bol, and if using the veterinarian version, it will be called Reforvit-B. (Due to the illegal nature of anabolic steroids, they will often be referred to interchangeably by the chemical name, the respective brand name, or the street name for that product. This can make it confusing for a beginner, but as we take a look at each of these steroids, you will get use to the various terminologies.) This was the first steroid to be used by athletes in the 1950's. Although effective, Dianabol can increase the level of estrogen in the body. For users of anabolic steroids, estrogen is the enemy. Estrogen is a female hormone that causes the growth of breast tissue in males (commonly referred to as "bitch tits" by bodybuilders), which many bodybuilders have suffered from due to their use of some anabolic steroids.

Dianabol has been shown in practice to increase both size and strength fairly well. The half-life of this compound is around four to seven hours, and the typical dose for bodybuilders is 25 mg to 100 mg per day, depending upon whether the bodybuilder is combining D-Bol together with another steroid ("stacking") or not.

Another type of AAS is referred to as Fluoxymesterone, Halotestin, or Stenox. This particular compound has been chemically altered to better withstand the liver's metabolizing mechanism. In other words, it can pass through the liver without being inactivated, so that it can produce the desired result. This was achieved by analyzing the compound. Without this alkalization, it would require a much higher concentration of the compound to get the same results. This type of AAS does not seem to act by binding with the AR.

In practice, fluoxymesterone is reputed to increase strength to a great extent. However, actual gains in muscle mass do not seem to be as dramatic. Clinical doses rage from 2.5 mg to 40 mg per day, but bodybuilders typically used between 30 and 80 mg per day. The half-life of the compound ranges between 9 and 10 hours, and it does not easily convert to estrogen in the body.

Oxandrolone (also referred to as oxandrolone powder or oxandrolona) is another type of AAS. Oxandrolone is another anabolic steroid that does not seem to convert to Estrogen in users, making it fairly popular.

Oxandrolone has a reputation for increasing strength and reducing fat, so most users are looking more for a toned physique than a bulky appearance. The half-life of oxandrolone is around nine hours and doses for bodybuilders range from 25 to 125 mg per day.

Stanozolol (aka Winstrol) is another AAS that does not bind well to the AR and which does not lead to increased level of estrogen. One of the benefits of stanozolol is its ability to prevent progesterone from binding to receptors. One of the downsides of many steroids is that they do not block progesterone from binding, which leads to water retention.

Stanozolol, in practice, can increase strength and causes some increase in muscle mass. Without water retention, these increases can look more impressive than the greater increases caused by steroids that do cause water retention. The half-life of stanozolol ranges between 7 and 15 hours, and the common dose for bodybuilders is between 25 to 100 mg per day.

Oxymetholone (Anadrol) is a compound that does not lead to increased estrogen, but it has been known to possess progestogenic properties, meaning it can lead to water retention.

In practice, oxymetholone can lead to large increases in strength and muscle mass, although the increase in mass may not be apparent, due to water retention. The half-life of oxymetholone is around 7 to 15 hours, and typical dosage for bodybuilders will be about 50 to 150 mg per day.

Methenolone Acetate and Methenolone Enanthate (Primobolan) is another compound without the estrogen side-effects common to anabolic steroids. This compound can either be taken orally in the form of Methenolone Acetate or injected in the form of Methenolone Enanthate.

In real use, Primobolan has been found to bind very well to the AR, and although increases in muscle mass are less dramatic, the lack of water retention make any gains far more obvious (it should be noted here that many users of anabolic steroids wrongly consider any gain in weight to be due to an increase in muscle mass, but in reality, this is mostly simply weight gained as a result of water retention. Therefore, any gains in mass caused by an AAS will be more obvious with less water retention). This AAS has a half-life of about 6 to 8 days. Taken orally in the form of the acetate version, the dose for bodybuilders will be around 500 to 1500 mg per week, and between 400 to 1000 mg per week with the injected enanthate version.

Testosterone Enanthate, Testosterone Cypionate, Testosterone Propionate and Suspension (commonly called "T") are versions of a testosterone compound that may in fact cause an increase in the production of estrogen. Due to its tendency to cause a lot of water retention and many other side effects compared to other anabolic steroids, the decision to use these types of steroid is one to be taken seriously.

"T" does bind very well to the AR, and can produce fantastic gains in muscle mass and in strength, provided that the dosages are high enough. The cypionate version has a half-life of around 8 days, while Enanthate falls just below that, and the half-life of cypionate is much shorter. In a suspension form, testosterone has a half-life of only about 10 to 100 minutes. Typical dosages of "T for bodybuilders range from 500 to 1000 mg per week. Nandrolone Decanoate and Nandrolone Laurate (often referred to as Deca) is another form that binds very with the AR without producing estrogen.

This steroid can produce decent increases in muscle mass and strength with little water retention, but higher doses can lead to water retention. The half-life of the decanoate version is 6 to 8 weeks and a little longer for the laurate version. Bodybuilders usually take about 300 to 800 mg per week.

Although this list should by no means be considered exhaustive, it is a fairly good representation of the different forms of anabolic steroids that are available on the legal and on the black market. Each has its pros and cons, and if you choose to use anabolic steroids, you should absolutely do your own homework and decide which type is right for you and your circumstances.

As previously mentioned, the only legitimate use for anabolic steroids in the eyes of the law is for medical purposes. However, the majority of users of anabolic steroids use them for performance enhancement purposes; therefore, that is the usage we will focus on in this book.

CHAPTER 2.

BENEFITS OF ANABOLIC STEROIDS

It seems that there are reports in the news every couple of months of some well-known sports figure who gets caught using anabolic steroids or some other performance enhancer and gets banned from their sport. Considering the money at stake in professional competition, this should not be surprising. However, other than a pure desire for financial reward, what causes these top-tier athletes to risk their careers by using this illegal substance?

In addition to the medical uses, which we have already discussed, there are three primary reasons why athletes at all levels may choose to use anabolic steroids. These are increased muscle mass, increased strength, and faster recovery time after an injury.

Muscle Mass

It is not difficult to guess that the average person who is using anabolic steroids will almost always achieve much better results when attempting to build muscle and improve performance then somebody who is not using anabolic steroids. This seems obvious. What may surprise you, however, is the degree of difference between users of steroids and non-users of steroids.

Due to the illegal nature of anabolic steroid use, finding reliable sources of information can be difficult. However, studies do exist. In a 10 week study, whose results were published in the New England Journal of Medicine titled *The Effects of Suraphysiologic Doses of Testosterone on Muscle Size in Norman Men*, a team of researchers wanted to compare the difference between users of anabolic steroids and non-users, when it came to muscle mass increase.

This study lasted 10 weeks and involved a group of 43 men of normal weight, ages 19-40. All of the participants had some degree of experience with weight training.

Researchers divided this group up into four smaller groups. The first group did not do any form of physical exercise and received no steroids or any other performance-enhancing substances. In other words, these were just normal men who were not working out at all. The second group also did not do any form of exercise, but they did receive a weekly dose of anabolic steroids. In other words, drug users who were not working out at all. The third group did take part in weight training but did not receive any performance-enhancing substances. In other words, normal guys working out. Finally, the fourth group consisted of men both doing weight training and receiving weekly doses of anabolic steroids. In other words, drug users working out. For the purpose of this study, researchers chose testosterone enanthate as the anabolic steroid to be issued to the two groups who were to receive a performance enhancer.

All groups were placed on the same diet, which was standardized based on each individual's body weight. However, researchers were careful to ensure that each man received the same relative amount of caloric intake, protein, and nutrients. The participants who were taking part in weight training, i.e. the third and fourth group, were placed on the same supervised workout regimen each week.

This means that the only difference between the groups within the study was the division between those who were working out and those who are not working out, as well as those who were receiving steroids and those who were not receiving steroids. Everything else was standardized.

The results were not terribly surprising. The first group, who did not exercise and did not receive any anabolic steroid, experienced no significant changes in the amount of muscle mass they possessed.

The second group, who did not exercise but did receive anabolic steroids, gained an average of 7 pounds of new muscle. This is particularly impressive, considering that this group was not engaged in weight training.

Group three, who did engage in exercise without receiving any anabolic steroids, was able to build an average of 4 pounds of new muscle.

Finally, the fourth group, who did exercise and did receive anabolic steroids, was able to build an average of about 13 pounds of new muscle over the 10 weeks of the study!

So, what can we learn from this study? First, unsurprisingly it revealed that if you are not exercising and are not using anabolic steroids, you are not likely to build any new muscle. Second, it showed that anabolic steroids can cause significant increases in muscle mass. Even though both groups who did engage in weight training were using identical workout routines and were on the same diet, the results for those using anabolic steroids were over 3 times greater than for those who were not using anabolic steroids. Finally, perhaps the most surprising finding of the study revealed that taking anabolic steroid injections, even without exercise, will lead to a significantly greater increase in muscle mass than exercising without anabolic steroids.

To emphasize, the guys who took steroids but did NO exercise built nearly twice as much new muscle mass as the guys who were weight training 3 times a week without taking steroids. Insane, right?

The study also revealed some other interesting findings. Several of the guys receiving the anabolic steroids did develop acne, while only one man receiving no steroids developed acne. Two other men receiving the drugs did report breast tenderness, but these were the only side effects noted.

We have all heard horror stories in the media about so-called "Roid Rage," claiming that people using anabolic steroids are unable to control their emotions and will fly off the handle with no provocation. One of the more interesting results of this study shows that there was *no difference in mood between any of the groups.* Even the participants in the group receiving anabolic steroid reported no significant changes in their mood or outlook on life, and their family and friends also reported no change.

Strength

Again, due to the illegal nature of anabolic steroids, it is difficult to find reliable research on how those compounds affect muscle strength. Fortunately, the study that we have already discussed also took muscle strength into account. Here, too, the results are similar.

Researchers measured the strength of each participant using two exercises - bench-press and heavy squats.

The first group, taking no drugs and doing no exercise, experienced no significant difference in the amount of weight they were able to lift over the 10-week period.

The second group, taking no drugs and participating in weight training three times per week, experienced an average of a 21% increase in the amount of weight they were able to squat, and a 10% increase in the amount of weight they were able to bench-press.

The third group, taking anabolic steroids but not working out, experienced nearly identical results to the second group.

Finally, the fourth group, working out and taking anabolic steroids, increased their muscle strength in the squatting exercise by about 38% on average and in the bench press exercise by around 22%. This represents nearly double the increase in strength experienced by those only working out or only taking anabolic steroids!

Healing Time

It is difficult to answer the question of to what degree steroids can reduce recovery time in a general way. This is because there are a variety of factors that can influence the outcome. For example, how much you weigh, how much muscle you have, what type of steroids you were using, and the severity of the injury. All of these factors play a part in determining how long it will take to recover from a muscle injury.

However, there is no question that the use of anabolic steroids can aid in recovering from a muscle injury.

To put it simply, steroids block the production of cortisol. The body produces cortisol when it is under stress. Cortisol can act as an anti-inflammatory and actually slows down the healing time of muscles, so blocking the production of cortisol can decrease your recovery time.

So, taking these three factors into account, it is easy to see how the use of anabolic steroids can benefit not only professional athletes but any man who wants to improve the quality of his physique.

Chapter 3.

Drawbacks Of Anabolic Steroids

With all of these proven benefits, it may be difficult to understand why anabolic steroids are illegal and are generally frowned upon. Therefore, we will take an in-depth look at the disadvantages associated with using anabolic steroids. Here, too, reliable information is hard to acquire due to the illegality of the substances, but we will discuss some of the most obvious drawbacks.

At the end of the day, it is your body, and if you choose to use anabolic steroids, you need to be fully informed of the potential risks involved. To make things simpler, we will divide the risks into three categories - societal risks, cosmetic risks and health risks.

Societal Risks

The first and most obvious drawback of anabolic steroids is legality. Although it is not terribly difficult to acquire anabolic steroids, the possession, use, and sale of anabolic steroids is illegal within the United States, except for prescribed medical purposes.

Professional athletes are regularly tested for performance-enhancing drugs, such as anabolic steroids, and can be banned from their sport if they are caught using them. From a competitive standpoint, this is quite understandable. The purpose of professional sport is to measure the athletic ability of the athlete, not to determine who can acquire the best anabolic steroid.

Another reason the use of anabolic steroids is frowned upon by society is the fear that users of anabolic steroid will obtain results that cannot be achieved by normal men simply by working out. These normal men, who do not know that the person, whose physique they're admiring and attempting to replicate, is using steroids, so they may injure themselves trying to work out hard enough to compete with those who are using anabolic steroids.

Another problem is that the greater muscle mass and leaner physique of the men using anabolic steroids may lead to unrealistic expectations from women, who expect that all men should look like that.

Finally, although the study that we discussed above did not note any instances of this, it is believed by many that using steroids will lead to uncontrollable emotional outbursts, also known as "roid rage."

Cosmetic Risks

Although anabolic steroids are primarily used to improve the user's cosmetic appearance, there can be side effects which must be considered.

We have already previously alluded to the first risk, gynecomastia, which is commonly known as "bitch tits." Numerous users of anabolic steroids who did not take the proper precautions found that they began to develop breast tissue. For many users of anabolic steroids this is a serious drawback, but with the proper precautions, it does not have to occur. Women experience similar risks when using anabolic steroids, in the form of increased facial hair.

Another risk of taking anabolic steroids is the development of acne in adults. This is a relatively common side effect of anabolic steroids. This is most likely caused by the greasy skin that many users of anabolic steroids experience.

Finally, an increase in the amount of testosterone present in the body can lead to hair loss and eventual baldness in men.

So, all of these downsides are real and present, but with the proper precautions, they can either be mitigated or totally eliminated.

Health Risks

The simple reason why most men choose to use anabolic steroids is to become healthier. Although the health benefits - including weight loss, increased muscle mass, increased vitality, and reduced healing time - are real and significant, there can be negative side effects of using anabolic steroids.

Liver damage. Anabolic steroids, no matter how they are taken, must eventually be processed by the liver. The liver is an organ, which is simply not designed to process such high levels of hormones. If a person is not careful and uses anabolic steroids for long enough, they may be at risk of developing liver cancer.

Because of the use of needles to inject anabolic steroids, some users of anabolic steroids can be at a higher risk for Hepatitis B, Hepatitis C, and other blood-borne diseases.

Due to users of anabolic steroids being stronger, they are at greater risk of tearing their tendons, because of the amount of weight they can lift.

Finally, your testicles will almost certainly shrink if you are using anabolic steroids. This is because, when you are using anabolic steroid, you are flooding your body with a substitute for naturally occurring testosterone. Because your body has such an excessive amount of testosterone flowing through it, your brain will send a message to your testicles to stop producing testosterone. This causes them to shrink, and in the long run, can even lead to atrophy. It is possible to avoid this shrinkage, and we will discuss the methods favored by many users of anabolic steroids to do so later, but this risk is real. After stopping the use of anabolic steroids, the testicles will usually revert to their original size, but not always.

Some of the other health risks associated with using anabolic steroids include:

- Nausea and Vomiting
- High Blood Pressure
- Joint Pain
- Jaundice or Liver Damage
- Urinary Trouble
- Heart Disease and Cancer

Now we can see that anabolic steroids are not a magical formula which will allow you to sculpt your body the way you want it without any consequences. However, most of the risks associated with using anabolic steroids are well-known and possible to prevent with the appropriate precautions.

Chapter 4.

How To Use Steroids

If you have thoroughly read and understood all of the potential risks and rewards associated with anabolic steroids and you still wish to use them, read on. We will now discuss the proper use of anabolic steroids.

For obvious legal reasons, we will not discuss how to obtain anabolic steroids here. However, even a cursory search on the internet should be sufficient to get you started. Another, usually more expensive option, is to simply find a drug dealer who specializes in steroids. If you choose this route to obtain your anabolic steroids, be aware that you may not always get what is advertised. Also, please do not infer from this that you should simply approach the largest person in a gym and ask them if they know where you can get any steroids. Anabolic steroids are illegal, so you should be as discreet as possible in your search for them.

Also, before you get started looking for anabolic steroids, you should know what kind you are looking for. Different types of anabolic steroids work upon the body in different ways, so before getting started, you should decide beforehand what results you are trying to achieve with your body.

With that out of the way, let's dive right into the actual use of anabolic steroids. Anabolic steroids are taken in "cycles." A cycle simply refers to the period of time that you are using anabolic androgenic steroids (AAS's) to improve your physical appearance. This improvement can refer to increasing muscle mass, increasing strength or decreasing the amount of fat in your body.

Cycles using oral steroids typically last about 4 weeks, and cycles using injectable steroids in combination with oral steroids can last up to 14 weeks. But, the length of your cycle will vary, depending on how your body responds, how advanced you are at using steroids and what specific gains you are looking to achieve.

The length of time you will stay on a cycle and the length of time between cycles mostly depends upon what you are hoping to achieve, as well as your comfort level with risk. If you wish to gain 35 pounds of lean muscle mass in only 2 months, the odds are good that you will have to spend the entire 2 months on your cycle without a break in-between, to achieve those results. If, however, you care about safety and your goals are a little more realistic, i.e. only gaining 8 to 10 pounds of muscle, then a cycle of two or three weeks "on," followed by four to six weeks "off" would be better.

In general, shorter cycles are safer, while riskier cycles are more effective. This is the dilemma you will face when using anabolic steroids - the most effective compounds are also the most dangerous. But, all steroids come with some level of risk. There is no such thing as a completely safe steroid. While one type of anabolic steroid may not be associated with bitch tits, it may harm your liver. And while another may not damage your liver, it could increase your risk of baldness. This is the trade-off you make when you are using anabolic steroids.

As previously mentioned, there are two ways to take anabolic steroids - orally or by injection. Both come with their own risks. Orally ingested anabolic steroids are far more likely to damage the liver, while injected anabolic steroids bring all the risks associated with injecting any substance. This can include infection at the injection site, blood borne illnesses, etc.

Those looking for quick results usually choose oral steroids. Oral steroids do not possess ester chains which have to be processed by the liver first to become active. Their half-lives are also usually shorter, which means that your system will flush them out faster. This is why many athletes choose oral steroids.

However, we are talking about anabolic steroids here, so these benefits also come with many drawbacks. First, when ingested, they must first pass through the digestive system and then the liver to reach the blood and achieve results. Your digestive system will destroy a large portion of the steroids, so you must take a larger dose enough to be left over to achieve your desired results.

Taking such a large dose of any substance can seriously damage your liver, which is a vital organ.

Bodybuilders typically prefer to inject steroids over oral steroids. This is because they do not share the same health risks. The chemical compounds required to make oral steroids are not present in injected steroids so your liver will not need to process those compounds.

Additionally, injected steroids do possess ester chains, so they take longer for the body to fully metabolize and their effects last longer.

Finally, since they do not pass through the digestive system, there are fewer side effects than with oral steroids.

However, with the longer half-life (the amount of time the substance will spend in your system) is longer with injected steroids, it is easier for athletes to get caught. Some athletes find a way around this by injecting steroids at the beginning of the cycle and switching to oral steroids once the cycle is nearly over.

Another downside of injected steroids versus orally ingested steroid is the pain factor. Some people have higher pain tolerance than others, but you are injecting a needle into your body so there may be pain. Many users attempt to minimize the pain by choosing injection sites, such as the glutes, that have fewer nerve endings than other parts of the body.

Most experienced users will rotate their injection sites rather than always injecting at the same site, but the best place to inject steroids is the upper outside quadrant of the buttocks, though you must be careful not to hit the pelvic bone.

As previously mentioned, you should never, I mean *never* inject steroids into your veins! Steroids are injected into the muscle. Anabolic steroids are mainly injected into one of 8 different muscle groups. These muscle groups include the glutes, biceps, triceps, lats, delts, pecs, calves, and traps. Because the muscle groups mentioned here each have their own corresponding muscles on the other side of the body, there are a total of 34 sites on the body where you can inject anabolic steroids.

Glutes: Select a spot in the upper outside quadrant of your buttock. The same corresponding spot can also be used on the other glute.

Biceps: Check into the center of the bicep, either on the inside or outside of the muscle.

Triceps: You can inject into any of 3 spots on the triceps - the outer head, the lower rear head, or the middle rear head. Inject into the center of the chosen area.

Lats: Find the outer edge of the muscle and inject into the center of that spot.

Delts: This muscle group has three possible injection sites, but the best one is the head of the lateral deltoid.

Pecs: When injecting into the pectoral muscles, you can choose between the upper inside, the middle inside, or lower outer.

Calves: Most users prefer not to inject into the calves, but if desired you can inject into the center of the outer or inner head.

Traps: There is one spot, on either side of the traps that you can inject; inject into the middle.

Although there are a lot of potential injection sites, the glutes and delts are preferred by experienced users. You should avoid injecting into the calves or traps because this can be fairly painful.

When injecting any substance into your body, your primary concern should be safety. Follow some common-sense rules when doing so, including **never** sharing needles, never re-using a needle, sterilizing all of your equipment, and always injecting into a clean site.

Now that you have read and understood the warnings let's take a quick look at how injecting steroids actually works. Remember, steroids need to be injected directly into the muscle, which means you will need a relatively long needle. When injecting into the glutes, a 1.5-inch needle will be ideal. You can use a one-inch needle if you want to, but this may take longer to inject yourself, and you may not be able to reach muscle fiber with a needle that short. So, be safe and get 1.5-inch needles.

After the glutes, the thigh is the second most common injection site. For most men, a one-inch needle will be fine for the thigh; there is usually less fat there for the needle to pierce.

So, now that you have everything ready and you have chosen your injection site take a few steps to make sure you are safe. First, wash your hands with hot, soapy water for at least 1 minute. Next, you will sterilize the injection site with rubbing alcohol. Go ahead and give the area a good scrub.

Now, you will hold the syringe like a dart that you are about to throw. Insert the needle all the way into your chosen injection site at a 90-degree angle; in other words, stick the needle straight in.

Once the needle is all the way in, slightly pull the plunger back to make sure no blood appears in the syringe. If you see blood, pull the needle out and try again, because you have hit a vein. If no blood appears, go ahead and begin to depress the plunger to inject your steroid. You don't need to force it; just push slowly and steadily until the liquid is gone. Trying to force the liquid into your muscle too fast is only going to cause you unnecessary pain.

When the liquid is all gone, go ahead and pull the needle out of the injection site. It is always a good idea to cover the injection site with an adhesive bandage or a cotton ball held in place with tape.

After injecting their steroids, experienced users massage the area thoroughly to spread the compound evenly between the muscle fibers and avoid lumps that can develop. When using oil-based anabolic steroids, it can also be helpful to hold your filled syringe under hot water to increase the viscosity of the oil and allow the anabolic steroids to flow more smoothly into the muscle.

That's it. You have now successfully injected yourself with anabolic steroids. Not too difficult, and if done correctly, not too painful. Your final step will be to safely dispose of your syringe in a spot where nobody can injure themselves with it. You will absolutely not use the same syringe, because it is now contaminated and in any case is most likely dull.

Chapter 5.

Stacking Anabolic Steroids

No matter what type of results you are hoping to achieve from anabolic steroids, it is recommended that you take two types which will work together through different mechanisms to achieve a synergistic effect. This process is commonly referred to as "stacking." For example, you will be able to achieve better results by stacking nandrolone with something like stanozolol instead of stacking it with oxandrolone, because both nandrolone and oxandrolone rely upon binding to the AR to achieve their results. Therefore, you should find a combination where one drug does bind to the AR, and the other one does not.

Some stacks that many athletes find useful include the following:

450 mg per week Nandrolone, along with 50 mg per day of stanozolol

450 mg per week Nandrolone, along with 50 mg per day of methandrostenolone

40 mg per day Oxandrolone, along with 50 mg per day of stanozolol

500 mg per week Testosterone enanthate, along with 50 mg stanozolol or methandrostenolone per day

500 mg per week Testosterone or nandrolone, with 50 mg oxymetholone per day

600 mg per week Methenolone, with 50 mg per day stanozolol

Looking at the first stack as an example, you would inject 450 mg of nandrolone on day one and then 6 to 8 days later, another 450 mg and so on. The stanozolol (or whatever oral steroid you choose to stack with) should be taken as regularly as possible to allow it to remain in the bloodstream over the course of the day.

This is also where it is vital to know the half-lives of your drugs. To achieve maximum results, you will want to know how much of that steroid is currently active in your body at any given time. The goal is to keep the amount of steroid acting upon your body at any given time consistent.

If you know the half-lives of the steroids you are taking, you can approximate how much of the drug is currently in your system. For example, if on day one you inject 450 mg of a steroid, when you inject another 450 mg on day seven or eight, there should be around 225 mg from the first injection still active in your body, bringing your new total active amount to 675 mg. Of course, this number will immediately begin to decline as your body processes more of the drug, but with a little knowledge about the half-lives of your drugs and some basic math, you'll have a pretty good idea how much of the drug is active in your system at any time.

Half-life can vary, depending on many factors, which is why most drug descriptions give only a range of time, such as 8 to 10 days. If you are a larger man, generally speaking, the half-life will be significantly shorter. Learning how your body affects the half-life of the anabolic steroids that you are taking will help you maximize your results.

No steroid is completely safe, but a few examples of the ones considered by many to be *safer* to get you started on your research usually include stanozolol, oxandrolone, fluoxymesterone, methandrostenolone, and nandrolone. There are more, but this should be enough to get you started as a beginner. There are many different kinds of anabolic steroids out there, as well as many products only *claiming* to be anabolic steroids. It may not be easy to find a reliable source of AAS, but it is crucial. After all, just like with any other product you purchase, you want to make sure that you are getting what is advertised. If you think you have found a good supplier, look for some reviews on bodybuilding forums; chances are good that someone else has used that supplier and can give you feedback about whether they had a good or bad experience.

Chapter 6.

What You Will Need

In addition to the anabolic steroids you wish to use, you will need some other items to be successful. These are things that you should already possess prior to taking your first dose of steroids. None of them are impossible to find, so if you choose not to obtain them, that may mean that you are entering into the world of anabolic steroids lightly, and have not fully thought through the risks and rewards of anabolic steroid use. Everything listed in here will either help you maximize your results or remain safe while doing so, so don't leave any of these out.

First, if you will inject your steroids, you will naturally need some syringes. Certain states have restrictions on buying syringes, but in most states, you can simply pick them up at your local pharmacy. If you are not comfortable purchasing them from the pharmacy, a brief search online will find you plenty of options to purchase them. They are not very expensive, and usually run about $0.50 apiece.

Remember, you should usually stick with a 1-inch to 1.5-inch syringe. If you can only choose one of these options, go ahead and pick up the 1.5-inch syringes so that you will certainly be able to inject yourself in the buttocks, which is the preferred site for steroids.

Depending on what sort of cycle you have in mind, you can decide how many syringes you will need to make all of your injections. Generally speaking, 10 syringes is the minimum requirement.

Next, you will get a bottle of rubbing alcohol, some cotton swabs to sterilize the injection site and probably some Band-Aids. No need to go crazy here; we are only making an injection, not performing open heart surgery so latex gloves will not be necessary unless you are injecting someone else.

You are also going to want to get some other supplies to maximize your results while minimizing your risk. Some of these may be pricey, but they will help avoid or mitigate many of the side effects caused by steroid use.

First, get a drug called tamoxifen (sold as Nolvadex), if you are using an aromatizable steroid (a steroid that causes increased estrogen production in men). Tamoxifen is a powerful drug that prevents receptors in the body from binding with the increased estrogen. If you are using an aromatizable steroid without using tamoxifen, you may end up with gynecomastia - bitch tits. This same goal can be accomplished with large doses of a drug called clomiphene, but we need the clomiphene for another purpose.

As mentioned, large doses of clomiphene can prevent the binding of increased estrogen with breast tissue, but the more common use for clomiphene when using steroids is to help you jumpstart the natural production of testosterone in your body after your cycle is over. If you do decide to use anabolic steroids, after your cycle is over, you will be a wreck. This is because your body has become accustomed to massive amounts of hormones flowing through it, so once your cycle is over and those hormones are no longer active in your body, you will feel sluggish and lazy. Taking a drug like clomiphene can reduce this malaise by causing your body to begin producing its own testosterone again. This will also help your balls regain their size, which is important in its own right.

During your cycle, you may want to take a drug known as clenbuterol. Clenbuterol is a fat burner that prevents water retention. As previously mentioned, one of the biggest side effects involved with the use of anabolic steroids is water retention. In fact, many users of steroids mistake the gaining of water weight for muscle mass increase. Taking clenbuterol will prevent this water retention and allow you to show off your newly acquired lean muscle mass.

After your cycle, you may also want to consider a drug called anastrozole. Anastrozole, sold as Arimidex, is a drug that can reduce bloating, decrease estrogen levels in the body, prevent joint pain, and increase testosterone production. Anastrozole can be fairly expensive compared to the other options listed here, though, and it functions mimic those of the other drugs listed, so this is more of an optional investment.

However, as mentioned, the tamoxifen, clenbuterol, and clomiphene should be considered just as essential as your syringes or the steroids themselves. Again, they may be expensive, but consider this an investment. This is your body, and many of the unpleasant side effects caused by taking anabolic steroids can be safely avoided by purchasing these drugs.

If you are taking non-aromatizable steroids, you can safely do away with the clomiphene, as there should be no excess estrogen produced, but you will still need the tamoxifen to help you rebuild your testosterone levels.

To avoid hair loss, you may also want to consider making an investment in finasteride (Propecia). This should also be considered an optional step, only to be done if you notice your hair starting to fall out.

Acne can be avoided by keeping your skin clean, using over-the-counter acne treatments, and by choosing a form of anabolic steroid that is less androgenic in nature.

Water retention can be avoided with a combination of reducing the amount of sodium in your diet, taking the drug clenbuterol, and avoiding aromatizable steroids.

To avoid liver damage, simply stay away from 17-AA steroids, and if you must use them, keep your usage as short as possible.

Will all of this be expensive? Probably. Prices will vary, depending upon your source, where you live, the quality of the product you are obtaining and countless other factors. But again, this is your body and now is not the time to skimp. You will be injecting or ingesting these drugs into your body, so you want to make sure you have a reliable source. A cheaper, or at least a more available option, may be steroids that are made for animals. Chemically, these are identical to those produced for humans, but since it is not illegal to give steroids to an animal, these may be easier to find, and hopefully cheaper.

Most of all, though, do your homework. There are many forums online where users will post their experiences with various drugs and suppliers, so read up on this. Also, eventually, you are likely to meet other people who share your interest in steroids, and these people may also be a great source of information for you. If you are serious about steroids and improving your physical performance, you can never learn too much.

In conclusion, we can see that most of the physical risks associated with taking anabolic steroids can either be mitigated or totally eliminated by using a little common sense, researching beforehand, so you know what you are doing, and having all of the necessary ingredients.

Sadly, however, all of the precautions in the world cannot eliminate the legal risks involved. Never forget that anabolic steroids are illegal and can bring about serious consequences, so if you choose to use them and get caught because you were not being discreet, don't say I didn't warn you.

Chapter 7.

Maximizing Your Gains

We have now discussed the use of anabolic pretty thoroughly. You should now have a fairly good idea of what you can hope to achieve using anabolic steroids, as well as the risks involved. In this final chapter, we will focus on what you can do while you are taking steroids to maximize your results.

Regardless of your reason for choosing to use anabolic steroids, or how long you been using them, all users share one thing in common: all users desire to maximize their results while limiting the side effects. Whether you are hoping to build lean muscle mass, to burn fat, or just because you love doing it, anabolic steroids most likely represent a fairly substantial financial investment for you. So, with that in mind, we will discuss some ways that you can get the most out of your investment and maximize your results.

In this book, we have learned many of the amazing things that steroids can do when it comes to improving your body. Unfortunately, steroids are not magic. When it comes to what you eat, the rules pretty much stay the same, whether you were using steroids or not; what you put in is what you get out.

Anabolic steroids do not perform miracles, they simply accelerate the process of growing muscle. So, if you are eating nothing but junk food, you will get fat. Also, building all of that new muscles so quickly for your body. So, proper nutrition is key to keeping your body working as efficiently as possible.

Whether you are taking anabolic steroids to bulk up or to cut some extra fat, following the proper diet will help you get the most from your investment. If you adjust your diet to match your goal for that cycle, that will help you stay strong and healthy and keep you feeling great all the way to the end.

The most important thing when taking anabolic steroid is to make sure that your body receives enough nutrients to keep it operating at Peak capacity. And this chapter, we will take a deeper look into how you can eat better to assist your body and making the changes you want from it. You can adjust the diet slightly based on whether you are bulking up or cutting down, but the foundation of a healthy diet will remain the same - proteins, carbohydrates, and fats.

Bulking

If your reason for taking anabolic steroids is to increase muscle mass, it is very important to keep a close eye on body fat. Because gaining weight usually goes together with adding fat, you will need to strictly regulate your daily caloric intake and keep your fat percentage under control. Beware, the more carbs and fats you eat, the more fat you will gain.

If you are a smaller guy who has always had trouble putting on weight, and you want to get into a cycle to bulk up, then you need to take in more calories than other body types will need. That's why there is no one-size-fits-all diet that I can give you; you will simply have to adjust the diet recommendations here to your own body type.

When bulking, proteins are the most important components of your diet. These are what allow your muscles to grow and to heal. This works by your body breaking protein from food down into amino acids and then transporting those amino acids to your muscle cells.

It is important to make sure that you are eating complete protein sources like eggs, meat, and milk to make sure that you have the correct ratio of essential and non-essential amino acids in your body. As mentioned, protein is extremely important for muscle growth. At a more basic level, however, it is actually the building-block - the amino acids- that need to be present in your body in the right proportions, to create an ideal anabolic environment.

The best way to achieve this is to eat enough complete proteins, but how can you know whether your diet is actually contributing to muscle growth, considering all the different variables involved, (body type, how quickly your body can process protein, the amount of rest you're getting, etc.), that can often cancel out your supposed optimal protein intake?

The good news is that protein status can be ascertained quite easily through a blood test to determine your nitrogen levels. Nitrogen testing is by far the most widely-used laboratory test to determine the body's anabolic state. Is shows whether nitrogen levels in the body are positive, (which you want); negative, (which you absolutely *do not* want); or neutral (which is the absolute minimum level you can work with).

A positive nitrogen level is the optimal state for growing muscle. Basically, when your nitrogen levels are positive, it means that you have achieved enough rest after your last workout, so the higher the nitrogen level in your body, the faster the workout recovery.

The average daily protein intake for those looking to add weight should be about 1.5 to 2 grams of protein for each pound of body weight. So, if you weigh 200 lbs., then your minimum daily intake of protein should be about 300 grams. If you're one of the guys mentioned above, who have a hard time putting on weight, you may need to raise that portion to about 400 grams of protein that you will carefully divide throughout the day. However, one gram of protein has 4 calories so bodybuilders can consume up to 1200 calories a day just in protein. These numbers are very important to remember when you are setting your daily caloric intake.

The best sources of lean, complete proteins include

Fish

Turkey breast

Chicken breast

Egg whites

Non-fat or low-fat dairy products

Shellfish

Lean red meats (top round, lean sirloin, and flank)

Carbohydrates are what your body uses to add more weight. Since carbs provide your body with energy, you will need to keep that level quite high to be able to train hard and retain your energy throughout the day. Also, they will keep your body from burning proteins for energy, and hold your blood sugar levels constant.

To make sure your blood sugar levels remain constant and you keep your energy up, you should eat carbs three times a day - in the morning and before and after your workout.

Your daily carb intake will be around 800-900 grams, and since each gram of carbs contains 4 calories, so that will add more than 3000 calories added to your daily intake.

Some great sources of carbohydrates include

Wheat bread

Pasta

Oats

Vegetables

Rice

When trying to add weight, you will need to eat about 100 grams of fat per day. You want to be sure that the fat you are eating omits unsaturated fats, such as Omega-3 and Omega-6.

The best way to get this fat intake is by eating fish. 100 grams of fat is what is required for bulking, so 100 grams of fish per day is recommended. Each gram of fish contains 9 calories so healthy fats will add around 900 calories to your daily intake.

So, the diet of someone trying to bulk up will usually consist of 30% protein, 50% carbs, and 20% fats.

A person who is looking to bulk up will need to consume between 4000 and 5300 calories per day. This equals 300 grams of protein or 1200 calories; 800 grams of carbs, or 3200 calories; and 100 grams of fat, or 900 calories. You may need to slowly adjust these numbered, depending upon your own body's requirements.

Cutting Weight

Now, if you want to get ripped, you will need to burn fat. This means you're going to have to decrease your daily caloric intake. However, this will not change what tools your body will need to achieve your goals; you will still rely on proteins, carbs, and fats.

Your protein consumption will have to be very high since you do not want to lose your lean muscle, you only want to lose your fat.

When cutting, you will need to reduce your carbohydrate intake. Carbs are necessary for energy and to replenish glycogen in your body, so you will still eat them in the morning, before your workout, and after your workout; you will just consume less.

Fats will still be very important to you when you are cutting since they contribute to the fat burning process. Just like when you are bulking, you will still look for unsaturated fats.

So, the diet of someone trying to cut fat will usually consist of 50% protein, 25% carbs, and 25% fats.

In the case of cutting fat, not only will you change the ratios of what you are eating, but you will also require a much lower daily caloric intake. When cutting, your daily intake should be between 2000- 2500 calories, most of them from proteins.

However, whether you were a bulking or cutting, it is important to schedule your meals throughout the whole day to keep your energy levels up. Instead of three meals a day, you should eat five or six smaller meals and stick to the schedule, wherever you are and whatever you are doing. You don't need to be hungry to eat, so get your body used to a diet routine, which will make sure it gets all the nutrients it needs for bulking or cutting.

Do not drink alcohol while cycling. Not only will this place added stress on your liver and kidneys, but it will also dehydrate you. Instead, drink as much water as you possibly can, at least 2 liters per day. That will keep you hydrated and make sure your kidneys are protected against the high amounts of proteins you are consuming.

Now, all of these numbers are just a starting point, since everybody is different and every body works differently. You will almost certainly need to slowly make adjustments to these numbers to find what works best for your individual needs.

As we have seen, anabolic steroids are capable of producing some amazing results. Your goal, if you choose to use anabolic steroids, is to make sure that your body is prepared for the dramatic changes you are asking for it. With a proper diet and the correct ratios of proteins, carbohydrates, and fats you can achieve your goals, whatever they may be.

Chapter 8.

Diary Of A User

This is my story of a year of using anabolic steroids. I had always been in and out of gyms for years. Never quite able to get a regular routine and because of this I was never getting good results. I thought to myself what's the point of all of this hard work if I'm barely getting anything out of it. I thought that I needed some help so I contacted a local personal trainer to help me with dieting and to train me once a week.

The trainer taught me so much about what I need to be eating, when I should be eating as well as helping me out in the gym. My trainer could easily lift twice as much I as could, more with a spotter. I had always wondered how can you possibly get that strong without being completely committed to working out on a professional level. After a few weeks of training with him he dropped the hint to me that he was using steroids. Curious by this I asked him more about it and in the end I asked him to get me some. I thought it would help me speed up the muscle growth, reach my goals faster and then I would get off it.

When I picked up the steroids I was so excited to start but it was mid week and I had to wait until Monday to start my cycle. I was so nervous when it came time to inject. I had no experience in doing anything like this. I watched videos on youtube on how to do it safely. I did it but it wasn't easy. I found that having a hot shower just before injecting made it much easier.

The first couple of days I didn't feel any different. I wasn't sure how long it would take to be fully in my system. I remember waking up one morning and muscles weren't sore. Normally whatever muscle group I just trained the day before would be feeling sore for a day or so. I felt like I was lifting more weight at the gym. My trainer even commented that I was getting stronger. I wasn't getting sore at all after training, that was probably one of my favorite things about using. Nobody likes going real heavy on legs one day and then barely be able to walk the next.

Training and lifting weights became my favorite part of the day. It was what I most looked forward to. I started growing, really fast. On average I was putting on around 2.5 lb (1kg) a week. I was feeling stronger than ever. New personal bests every week. The diet was pretty bland and boring but I had 2 cheat meals every week which I looked forward to.

I completed 3 cycles within in a year. Each cycle was 12 weeks long with a 6 week break in between. During the 6 week break I also did a 2 week clenbuterol cycle. I initially just wanted to build muscle and bulk up but with the proper diet to go along with the steroids I was capable of reducing body fat easily while still getting good muscle growth. I remember my friends couldn't believe their eyes. I didn't tell them I was using steroids. They all wanted to train with me and learn my secrets.

Cycles

Cycle 1

Sustanon 250

100 Dianabol Tablets

Anastrozole Tablets

This first cycle I injected 1 ml (250mg) of sustanon 250 every Monday and every Thursday for 12 weeks. My injection site was my glute alternating each time.

I had 3 dianabol tablets everyday breakfast, lunch and dinner until I ran out.

Anastrozole tablets are to reduce the extra estrogen levels in your body from using steroids. I had half of a 1 mg tablet every second day.

Cycle 2 And 3

Deca Durabolin

Test Enanthate

100 Dianabol Tablets

Anastrozole Tablets

This second cycle I injected 1 ml (250mg) of deca durabolin and test enanthate every second day for 12 weeks. I was still injecting into my glute alternating each time. As for the dianabol tablets and anastrozole tablets it is the same as the last cycle. A dianabol with breakfast lunch and dinner and half a anastrozole tablet on injection days.

DIET

This is the diet I used throughout my 3 cycles. It worked very well for me but may not be ideal for everyone. There may need to be tweaks here and there such as the quantities increased or decreased. I'm not a fan of fish so there is a lot of chicken in this diet but feel free to substitute some chicken for fish.

This diet is roughly 3100 calories per day. The key is to not let yourself be hungry. You should be eating every 2-3 hours. If you finish your meal and find you are still hungry, that's an indication that you need to increase your portions. Eat as many green vegetables as you like, the more the better and drink 3-5 liters of water per day.

I had 1 cheat meal every Wednesday and Sunday night. It helped break up the week and gave me something to look forward to. For the cheat meal give yourself an hour of eating as much as you can or whatever you like. The idea of the cheat meal is to spike your metabolism. After eating a clean diet for a few days your metabolism will slow down and relax because it doesn't have to work so hard. Cheat meals are not needed if you don't like but it made it easier for me.

Meal 1

3 eggs

7 oz or 200 grams of lean red meat (steak, venison, kangaroo)

Green vegetables

500ml of water

Meal 2

7 oz or 200 grams of chicken breast

6 oz or 180 grams of basmati rice

Salad (lettuce, cucumber, capsicum, etc)

Meal 3

6 oz or 180 grams chicken breast or 2 tins of tuna (if you have tuna add in 20 grams of almonds)

150 grams basmati rice

Green vegetables

Meal 4 (Pre Workout Meal Approx 1 Hour Before Training)

1 serve of whey protein

2.5 oz or 75 grams of oats

Intra Workout

1 serve of intra workout (BCAA)

1 serve of glyco load (high carb supplement)

Post Workout

1 serve of creatine

1 serve of whey protein

Meal 5

7 oz or 200 grams of chicken breast

9.5 ox or 270 grams of sweet potato

BEFORE:

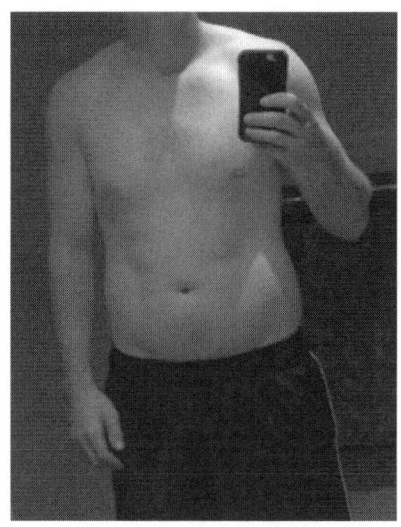

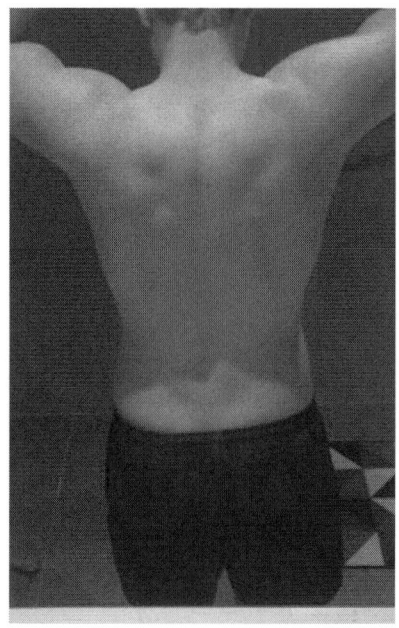

AFTER:

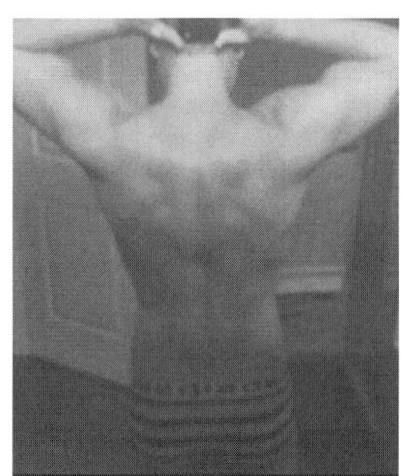

Conclusion

Thanks again for taking the time to buy this book!

You should now have a good understanding of anabolic steroids. Anabolic steroids have been shown repeatedly to cause massive improvements in physical performance and can greatly improve quality of life for those suffering from certain diseases. But, this does not come without risk. From being banned from sports to being thrown in jail, to serious health risks - these physical improvements do not come without a price.

That being said, with the proper precautions most of the risks associated with anabolic steroid use can either be minimized or avoided altogether.

This book was not written to convince you to use anabolic steroids or to not use anabolic steroids. As an adult, you own your body, and you are fully responsible for anything you choose to do with your body. The goal of this book has been to provide you with the information you may need to decide whether *you* and only you, wish to use anabolic steroids or not. It is also a guide for those who ultimately decide that they would like to take advantage of the physical benefits provided by anabolic steroids.

However, this is by no means a comprehensive guide. Although we have discussed most of the benefits and risks associated with using anabolic steroids, as well as how to avoid those risks, there should just be the beginning of your journey. As we have seen, there are numerous ways to maximize your results from using anabolic steroids, such as balancing your diet and keeping an eye on your caloric intake, we have barely scratched the surface. Hell, we haven't even talked about your actual workout. Although results can be achieved from steroids without working out, the *real* results come from putting in sweat at the gym. So, if you decide you want to use anabolic steroids, do your homework. From internet forums to fellow enthusiasts at the gym; the list of resources that can help you achieve your physical goals is endless. You have taken the most important first step by buying this book – getting started. Now, the rest of the journey is up to you!

If you enjoyed this book, please take the time to leave me a review on Amazon. I appreciate your honest feedback, and it really helps me to continue producing high quality books.

Simply CLICK HERE to leave a review, or click on the link: (Insert link here).

Printed in Great
Britain
by Amazon